A plain-English building architecture guide for those who wish to draw their own Vastu-compliant house plans

Vāstu Shastra Explained

Written and illustrated by
V. Subhash

Copyright
© 2020 V. Subhash. All rights reserved.

Based primarily on the *Vastu Shastra* found in *Matsya Purana*. A free PDF ebook download of the relevant extract from *Matsya Purana* is available at: **www.VSubhash.in**

First edition
Published in 2020 by V. Subhash

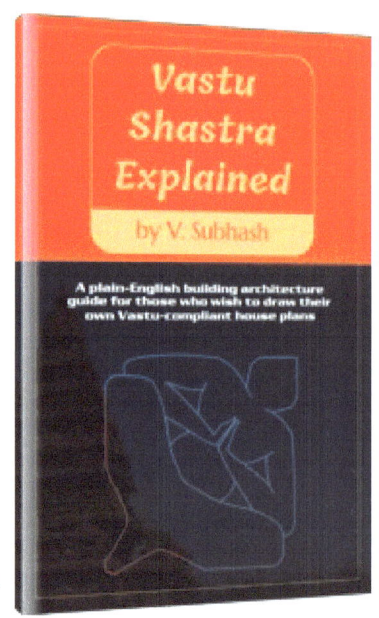

Acknowledgements

- All illustrations in this book are by V. Subhash except for images by artists from Unsplash.com — Jake Hinds (cover), Alberto Castillo Q., Serge Fedynyak, Francesca Tosolini, Janko Ferlič, Jean-Philippe Delberghe, Spencer Davis, Mitul Gajera, Jamie Mejias and Yash Shah.
- Brondell, Inc. (for photo of their product)

Preface

Most of those who build a house will do so only once in their lifetime. For this reason, it needs to be done right the first time. Even if you do not believe in age-old traditions like Vāstu, you might still want to follow its recommendations just for the peace of mind. To me, Vāstu seems to be a collection of time-tested architectural best practices that are part of our rich heritage. In this book, I explain the basic tenets of Vāstu Śastra straight from the text in *Matsya Purana*. I also provide information that would enable you to develop your initial house plan as per Vāstu Śastra. I have provided some sample house plans to explain basic Vāstu implementations. I have used the spelling 'vastu' although it is pronounced 'vaasthu'. The spelling 'vaastu' is also popular.

Disclaimer

Anything subject to interpretation has associated risks. I am not a Vāstu expert. I did some research for my own needs and wrote this book to record my findings. I read the Vāstu Śastra abstract in *Matsya Purana*. I read several books by modern Vāstu experts. I read translations of a few works similar to Vāstu Śastra. I wrote this book in the hope that it will be useful. The reader needs to do due diligence in all matters, not rely solely on any one book. Compliance to local building codes is very important. Vāstu or this book should not be blamed for any deviations. Most 'Vāstu experts' are professional astrologers. Some of them are 'consultants' of the Chinese *Feng Shui* (and see Vāstu Śastra in similar terms although the two are very much unlike). Anyone proficient in Sanskrit and can read ancient texts can become a Vāstu expert. The real Vāstu experts come from the many *mestri* or building contractor castes in India, although not many practice the art anymore. There are of course countless 'Vāstu experts' who upsell the concept to solve all kinds of problems (totally unrelated to buildings) with their 'Vāstu remedies'. I proffer no such thing in this book. Vāstu is about architectural efficiency and compatibility with natural elements. Expect nothing more from this book. I cannot provide any Vāstu consultation as it is not my occupation. You will need to see a local consultant. They are usually listed in the phonebook or in newspaper classifieds. However, remember my advice. Do not get bamboozled by any of them. If you have any doubts, read the English translation of Vāstu Śastra abstract in Matsya Purana. It is available on my website free of cost. You be the ultimate judge.

V. Subhash
22-7-2020
www.VSUBHASH.in

Contents

- Sources of *Vāstu Śastra* [5]
  - History of *Vaastu Devata*
- Vāstu Śastra recommendations [6]
  - *Vāstu Mandala* chart
  - Main entrance
  - Rooms
  - Trees
- Other *Vāstu Śastra* considerations [12]
  - Five elements
  - Room dimensions
  - Central area
  - Sides facing streets
  - Plot shape
  - Plot location
  - Plot orientation
- Modern Vāstu interpretations [17]
  - *Kubera* position
  - Water sources
  - Workplace
  - South-West corner
  - Staircases
  - Slope of floors and pipes
  - Bathroom and toilets
  - Septic tank
  - Ambiguous/conflicting interpretations
- Non-Vāstu considerations [22]
- Sample Vastu plans
  - House plans for East-facing plot [24]
  - House plans for North-facing plot [26]
  - House plans for West-facing plot [27]
  - House plans for South-facing plot [29]
  - A note on *naalukettu* houses [31]
  - Two-storey tiled house with full-height *thattil* [32]
  - North-facing plot [33]
  - South-facing plot [34]
  - East-facing plot [35]
  - West-facing plot [36]
- Annexures: Magnetic declination maps [37]

Sources of Vāstu Śastra

India has several ancient treatises on building construction. One of them is *Vāstu Śastra*. In fact, there are several *Vāstu Śastras*. They prescribe how buildings including houses, temples and palaces need to be built. They are all built around the belief of a *Vāstu Bhagawan* or *Vāstu Devata*. Vāstu Śastra is referred in many *puranas* and other ancient texts. You will find an abridged form in the 252nd chapter of *Matsya Purana*, the history of *Lord Vishnu*'s first avatar.

History of Vaastu Devata

A *deva vastu* (godly life form) was created from the sweat of Lord *Shiva* during a fight with an *asura* (demon). This vastu became so powerful and hungry that he threatened to consume all three worlds. So, the lords, rishis and even demons came together along with Lord Brahma to pin him down on Earth. The *deva vastu* protested this treatment and (as a compromise) gods allowed him to be worshipped as "Vaastu Devata" by those who create houses, temples and other buildings. Thus, the vastu became *Vaastu Devata* or *Vaastu Purusha*.

- English translation of Matsya Purana at **Kamakoti.org**
- The *Vastu Vidya* chapters of the *Matsya Purana* are available as a free PDF download from my website at **www.VSubhash.in**.

Vāstu Śastra recommendations

The *Vāstu Śastra* abstract found in **Matsya Purana** is brief enough and complete enough for ordinary people to make their own house plans. Here is what I found:

Vāstu Mandala chart

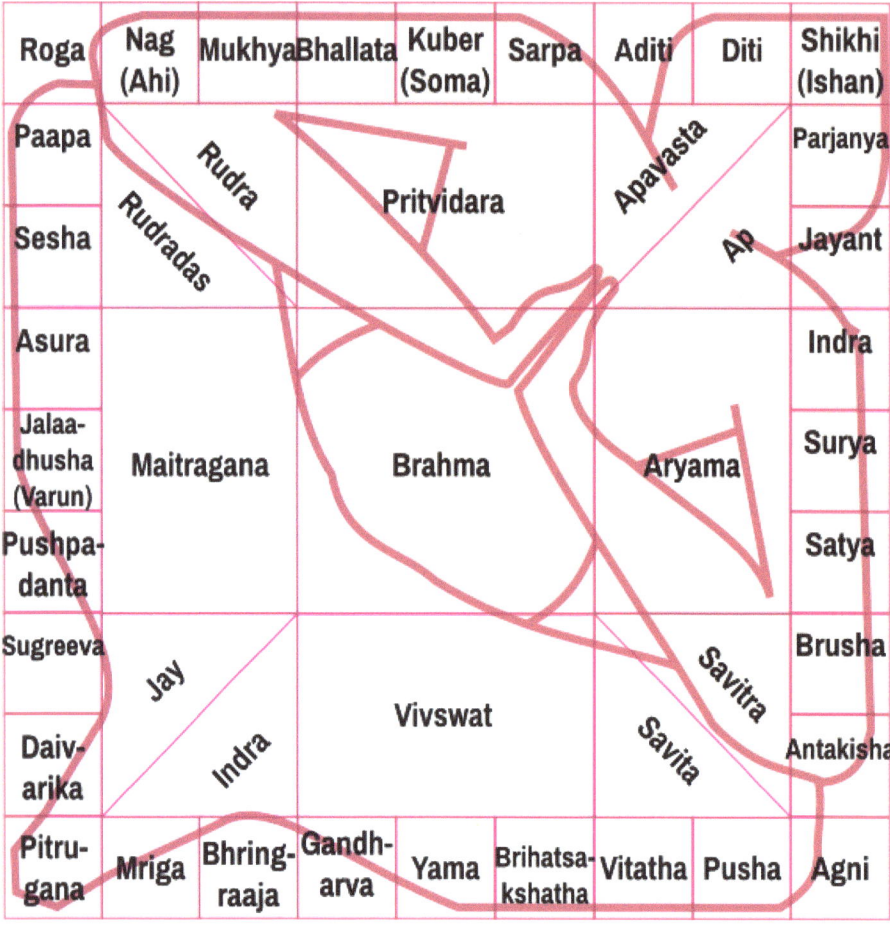

The layout of the house plan needs to divided into 81 squares. Thirty two devatas (*Baahya Devatas*) in the outer row of this "Vāstu Mandala" are: *Shikhi (Ishan), Parjanya, Jayant, Indra, Surya, Satya, Brusha, Antakisha, Vayu (Agni), Pusha, Vitatha, Brihatsakshatha, Yama, Gandharva, Bhringraaja, Mriga, Pitrugana, Daivarika, Sugreeva, Pushpadanta, Jalaadhusha, Asura, Sesha, Paapa, Roga, Ahi, Mukhya, Bhallata, Soma, Sarpa, Aditi and Diti.* At the centre is Lord *Brahma* (God of Creation).

The main implications of this chart are:

- The location of these devatas (gods) determine the ideal location for the rooms in your house plan.
- Because the *Vāstu Mandala* is a square, the overall house plan has to be a square or at least a rectangle. Anything else, such as circular or star-shaped plans, are not considered. Significant extensions distorting the square or rectangular shape of the plan are also not considered.
- The square or rectangular foundation has to be aligned to one of the cardinal directions - North, South, East or West - to get the house evenly exposed to the Sun.

- In plots that are diagonal to the main directions, the rooms would merely change their order along the sides so as to maintain their alignment with devataas in the Vāstu Mandala.
- No rooms are built in the centre where Lord Brahma is holding down the *Vāstu Devata*.

Main entrance

Main entrance can be on any of the four sides.

	Main door/entrance	
Main Door Faces	**Mandala Position**	**Devata**
East	6, 7	Indra, Jayant
North	6, 7	Bhallat, Sowmya
West	5, 6	Varuna, Pushpa
South	5, 7	Yama, Vitatha

- Main entrance HAS to be in the right half of the front wall, not in the

middle.
- It is a popular misconception that south-side entrance implicates death and misery as Yama (God of Death) position is located there. Vāstu Śastra recommends it as South is one of the cardinal directions. The only prohibition is having a door above a door (in another floor), which is considered as an invitation for *Yama* (Lord of Death). It probably weakens the structure.
- East-facing house is considered the best.

Rooms

The following rooms are recommended in these quadrants and sides:

Location of Rooms

Corner/quadrant/side	Room	Lord
North-East	Pooja room	Shiva
North	Water storage	Indra
South-East	Kitchen	Agni
North-West	High-value storage	Vayavya
South-West	Master bedroom; heavy material	Nairuti

Vastu Shastra expects South and West sides with more heft than the North or the East. Less free space is left on the South and West sides. (Separately, town-planning rules may mandate certain minimum space between compound wall and plinth area of the house.)

The land is expected to be sloping from West to East and South to North so that water flows in that direction.

Pooja room

The North-East corner (Shiki/Ishaan) is at the head of Vāstu bhagwan in the mandala chart. Vāstu Shastra recommends that the pooja room or place of worship be located here. Student's study or library could be placed in this quadrant.

It is best if the room encompassing this corner not used for a bedroom, although this may not always be possible due to paucity of space. If not avoidable, the pooja place should still be in North-East corner.

If a separate pooja room cannot be accommodated, Vastu experts recommend a pooja corner in the kitchen in the North-East corner.

Kitchen

The kitchen is recommended in the South-East quadrant with the stove to be placed in the South-East corner where Agni (Fire) is located. When South-East corner is not available, Vāstu experts suggest the North-West as an alternative, as it is ruled by Vayu (Air). In all traditional houses that I have seen, the kitchen was always in the Eastern side. So, I do not recommend using North-West for the kitchen. In the last chapter, my South-facing plot sample plan shows how to handle the situation where the South-East corner is not available.

Bedroom

- The master bedroom should be in the South-West quadrant.
- Other married couples can have bedrooms in the North-West quadrant.
- Bedrooms in the North-West quadrant is recommended only for very old members of the family.
- Bedrooms in the East is recommended for young members of the family.
- Bedrooms in the South-East quadrant are not recommended.

Possible mandala positions are given in the sample plans.

While sleeping, the legs should not be towards South. (You will wake up to see Lord Yama, the God of Death!) So, sleep with your head in any cardinal direction except North.

Vastu experts suggest placing heavy beds or an almirah the South-West bedroom to pin down the Nairuthi demon and because the South and West sides are recommended for heavy stuff. Do not store valuables or place a water sink or bathroom in the South-West corner.

Workshops, bathrooms/toilets and animal enclosures

Vāstu Śastra clearly recommends that workshops, bathrooms (and by implication, toilets too) and animal shelters be located outside the house.

In a later chapter, you will learn how these prohibitions are circumvented now.

Trees

Thorny and milk-yielding (Ksheera Vrikshas) trees are advised against. No trees are allowed close to the house.

Around the house, you can plant:
- Vata vriksha (Aal/ Aalamaram/Banyan/Ficus Benghalensis) in the East,
- Udumbara (Aththi/Gular/Indian Fig Tree/Ficus racemosa) in the South,
- Pippala (Aryaal/Arasamaram/Peepal/Ficus religiosa) in the West and
- Plaksha tree (some other type of fig) in the North.

Personally, I would not recommend any of the fig species on small plots, as they are epicytes.

Other recommended trees (for Lakshmi/providence) are:
- Pumnaga (Nagachembakam/Iravam/Nagakesara/Ironwood/Indian Rose Chestnut/Mesua ferrea),
- Ashoka (Saraca asoca),

- Shami (Vanni/Prosopis cineraria),
- Tilaka (Arni/Agnimantha/Clerodendrum phlomidis),
- Champa (Shenbagam/senbagam/Plumeria),
- Dadimi (Maadulampazham/Pomegranate),
- Peepali (Pippali/Thippiliver/Indian Long Pepper/Piper longum),
- Draksha (grapes),
- Arjuna (Neer maruthu/Marutham pattai/Terminalia arjuna),
- Jambeera (lemon),
- Puga (betel nut),
- Panasa (jackfruit),
- Ketaki (Kaitha/Kaithai/fragrant screw pine/Pandanus odorifer) (advised against Lord Shiva),
- Malati (Aganosma heynei),
- Mallika (jasmine), and
- Coconut, Kela (banana).

Other recommendations

- Construction beginning in certain months have certain benefits:
    - Vaisakha — 'gets cows and gems'
    - Ashadha — 'gets good servants, gems and domestic animals'
    - Shravana — 'gets good servants'
    - Kartika — 'gets wealth'
    - Margasira — 'gets plenty of grains and eatables'
    - Phalguna — 'gets a son and gold'
- Beginning construction in Chaitra, Jyestha, Bhadra and Ashwin months are advised against.
- For time and dates for rituals and construction activities, it is best to consult a priest or astrologer, rather than Vastu Shastra.
- Sunday and Tuesday are off-days for construction.
- The main door should be bigger than others but not too big.
- The South should be on higher ground compared to North.
- There should be more space in front than in the rear of the house.

FYI: Hindu calendar

Lunar month	Solar month	Western dates
Meena/Meenam	Chaitra/Chitirai	13 March → 12 April
Mesha/Medam	Vaisakha/Vaikasi	13 April → 13 May
Vrishaba/Edavam	Jyestha/Aanni	14 May → 14 June
Mithuna/Mithunam	Ashada/Aadi	15 June → 15 July
Karkata/Karkadam	Shravana/Aavani	16 July → 15 August
Simha/Chingam	Bhadra/Purataasi	16 August → 16 September

Kanya/Kanni	Ashwin/Aipasi	17 September → 16 October
Tula/Thulam	Kaartika/Kaartikai	17 October → 14 November
Vrishchika/Vrischikam	Margasira/Maargazhi	15 November → 15 December
Dhanu	Pausha/Thai	16 December → 12 January
Makara/Makaram	Megha/Maasi	13 January → 11 February
Kumbha/Kumbham	Phalguna/Panguni	12 February → 12 March

Other Vāstu Śastra Considerations

Vastu Shastra is not the be-all and know-all of building construction theory. It has to be applied with other knowledge we have learned over time.

Five elements

The "pancha maha bhutas" (Vayu, Jal, Prithvi, Agni, Akash) are divided along the main directions. As the Sun rises in the East and sets in the West, the location of various rooms should be considered in relation to Sun rays falling on the house.

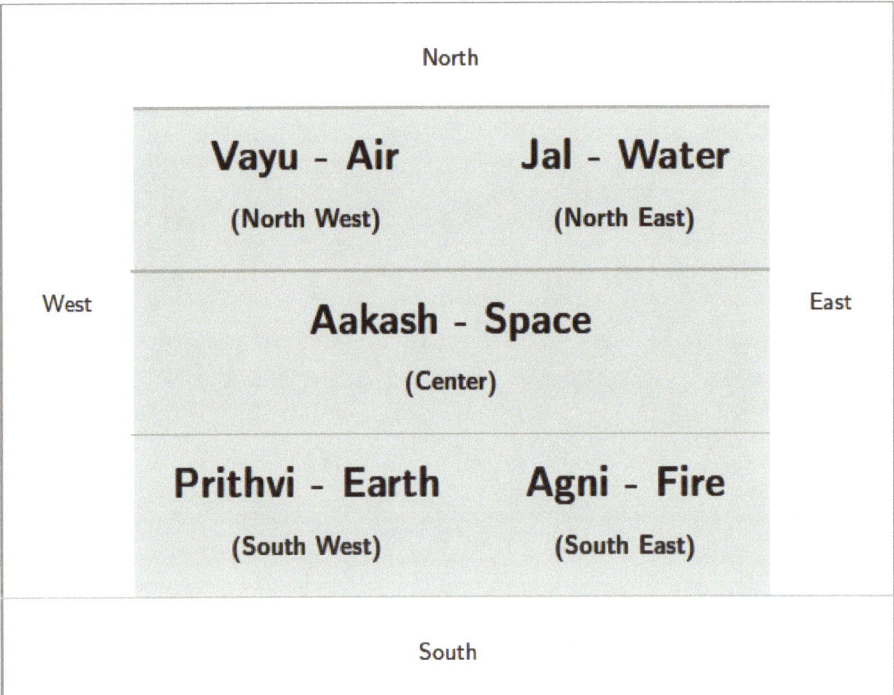

Room dimensions

Apart from Matsya Purana, commentaries by Vāstu experts today refer to other ancient works such as *Thachchu Shastram, Maya Mata, Brihat Samhita, Manasara* and *Manayadi Shastram*.

Old Tamil treatises prescribe the best lengths (in feet) for room sides as 6, 8, 10, 11, 17, 20+. I am not sure if an exact translation of the current international standard for a feet was made to the measurement referred by the Tamil texts. Most old plan documents use *hastas* or cubits. A cubit refers to the forearm length, from the tip of the middle finger to the elbow. A cubit is equal to 18 inches or 72 centimetres or 12 angulas.

Central area

The centre of the house (ruled by Brahma) should not used as a room. Ideally, it is used as the place where passageways meet within the house. This is better illustrated in the sample house plans available in the last chapter.

Sides facing streets

Plots with streets running parallel to the sides are good in the following order:

- All four sides
- East
- East and South
- South
- South and West
- West
- West and North
- North
- North and North East

Vastu authorities differ on plots with streets on three sides. In practice, for a residential plot, more streets means more dust and disturbance. For a shop or business establishment, 'the more the merrier' is true.

Plots that have streets end on any side can be a mixed bag. Streets can end on the left and right of each side. This is referred as *street focus* by Vastu experts today. Streets can also end in the corners. This makes 12 directions in which streets can end on a plot. Of them, a few are considered bad (marked with dotted lines).

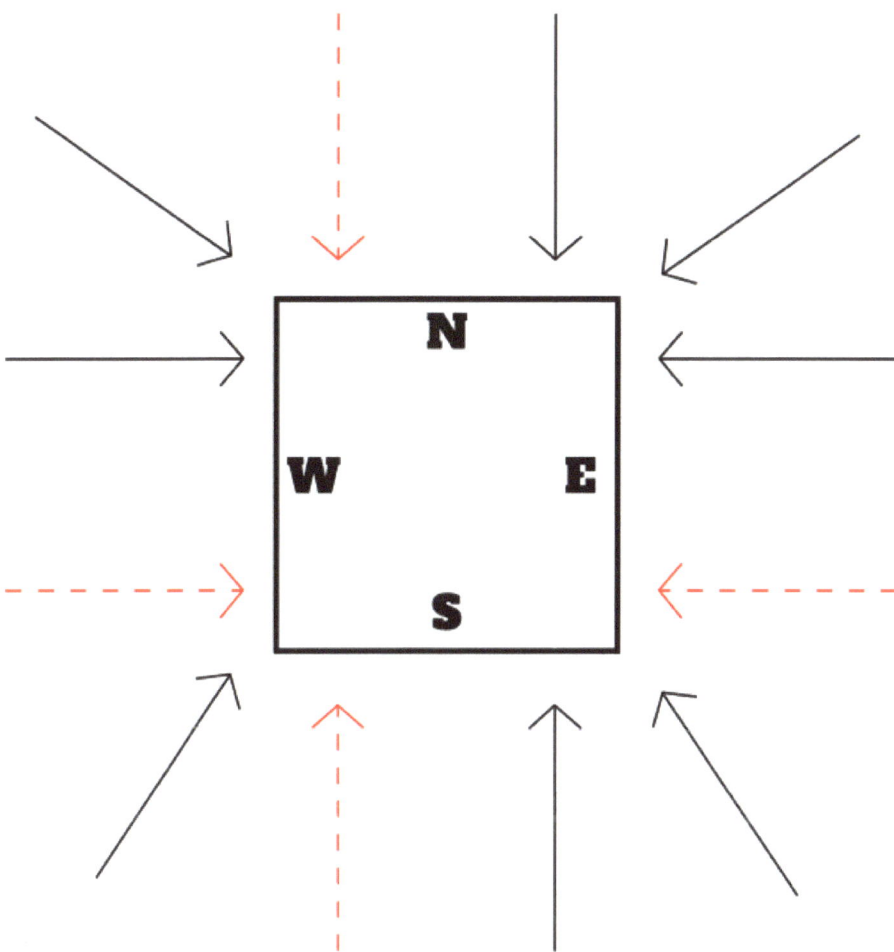

Plot shape

Like the house shape, the plot shape is recommended to be a square or rectangle. If it is not in that shape, you could carve a square or rectangle subplot inside this area after leaving some space all around it. You could have a wall fence for the full plot and a wire fence for the inner subplot.

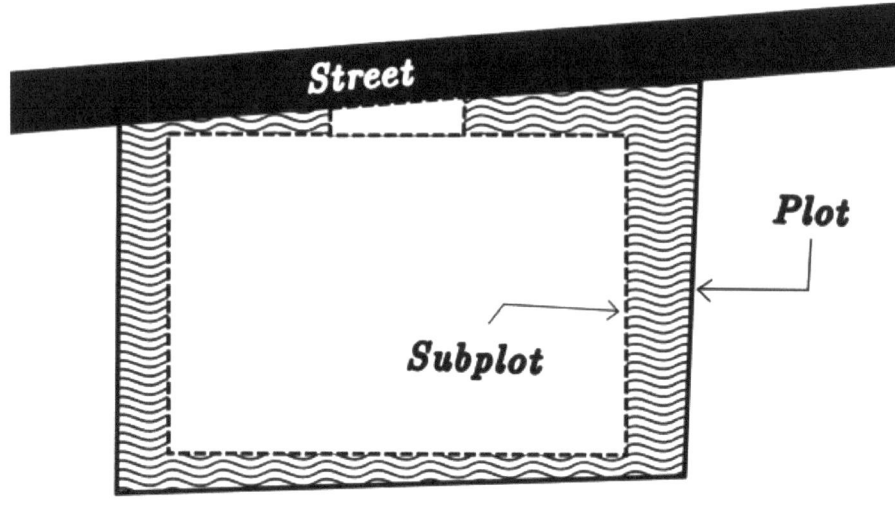

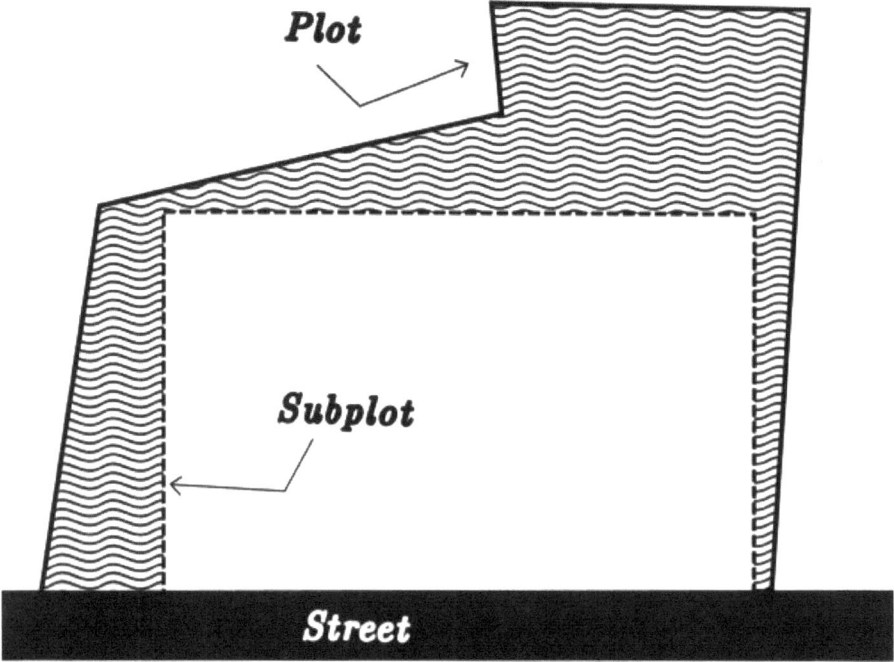

It does not matter if you lose some valuable space for your house. You can have a garden or use the space for storing materials, parking or animal shelters. Get the subplot in proper shape and build your house inside the subplot.

Plot location

Vastu Shastra considers it inauspicious when the shadow of a temple, hill or tree falling on the house. Within the plot, the prohibition is on trees near the house.

Plot orientation

When determining the orientation of a plot with the cardinal directions, you will need to use a magnetic compass. In some parts of the world, the directions shown by the magnetic compass is erroneous by a few to several degrees. To be sure, you need to consult a *magnetic declination map* [32] for correcting the compass error in your area.

Use the online calculator at this URL to determine the magnetic declination of your location and correct the directions shown by your compass. The error value can be off to the East or West. Eastern values need to be deducted. Western values need to be added.

https://www.ngdc.noaa.gov/geomag/calculators/magcalc.shtml

The geographical North and South (true North and true South) might be different from what is indicated by a compass. It changes over time. This discrepancy known as magnetic declination. In India, magnetic declination is insignificant for the length and breadth of the country. Most mobile phone compass apps automatically adjust for magnetic declination once you activate the location tracking (GPS, Glonass…). Professional models of magnetic compasses will have an extra bezel that can be rotated for eliminating the declination error but they will need to use it with a magnetic declination map. Magnetic declination maps with more resolution for your location can be downloaded and printed from:

https://www.ngdc.noaa.gov/geomag/WMM/image.shtml

In plots that are diagonal to the main directions, the rooms should merely change their order along the sides so as to maintain their alignment with devataas in the Vāstu Mandala.

Modern Vāstu interpretations

Vāstu Śastra is a science. It cannot be followed like how an orthodox fundamentalist pursues a holy book. It needs to be adapted to the time that people live in. You can seek to meet the broad recommendations - 100% Vāstu compliance should not be your goal, as it may not be practical or economical.

Kubera placement

Kubera (Soma) is the God of Wealth was an option for the study room. His position is now an option for main entrance, workplace and safe location for money and jewellery.

Water sources (wells and borewells)

Today, water sources need not be separate from fire places as wood or coal is not the main kitchen fuel. North, North-East and East are recommended for wells and borewells. For practical reasons, borewells will have to be located on the side where it is accessible by the road. (There may not be enough space around your house accessible by a borewell lorry.) Water tanks can be built in the West as it is Varuna bahawan's (rain god's) side. South-West is also good but not the South-West corner.

Workplace

Although karyalaya (workplace), bathroom, animal shelters and toilets are recommended outside the house, it may not be possible in urban places where space is limited. So, what was earlier recommended for a study room (Kubera or Soma) can now be used for a workplace.

South-West corner

In the Vāstu Mandala chart, the corners are occupied by four adjacent squares forming a larger square. These corners are considered "strength area" and are not recommended for the location of doors or windows. On the plot, all corners, except the South West, should be left empty or uncovered. Timber and/or leftover construction material is recommended to be placed in the South West corner, as it is the position of the Nairuthi demon. In all rooms, it should be the location of almirah and material stores (except food, water and wealth stores).

Staircases

"Heavy" structures such as staircases are recommended in the South West side. With staircases, the requirement is that the person should be able to descend facing East or North. Inside the house, one flight of steps is usually enough to connect a floor with the next one. Most Vāstu experts do not recommend winding or arched staircases.

Verandahs and balconies

Verandahs are open space around the house that is exposed to the outside. They are part of the foundation and need to follow the same Vastu rules. In summer, wide verandahs can protect the walls of living spaces from the rays of the Sun. You can build them all around the house.

Balconies are constructed in the upper floors and probably do not need any Vastu interpretation. Some Vastu experts prohibit them in places where the *street focus* is prohibited, For practical reasons, do not construct them over doorways of the lower floor. Apartment complex builders should build them in a staggered fashion across levels.

Apartment complexes

Every dwelling unit in an apartment complex should follow the same rules as a house built on the ground.

Slope of floors and pipes

The slope of the floors and the alignment of outgoing water pipes should be towards East or North.

Bathroom and toilets

Vāstu Śastra explicitly states that baths (and by implication toilets) be outside the house. Very few people can afford this luxury now so Vāstu has now been co-opted to find the best place for indoor plumbing.

If toilets have to be within the house, they are now recommended in the position below and right of the North West corner, closer to the outer walls with maximum supply of air and sunlight. Although two other locations in the East are permitted by Vāstu experts, I do not recommend them. I believe toilets need to be as far away as from the pooja room, kitchen, hall or the main living area, which are usually in the East.

Vāstu authors recommend that the commode should be placed so that the person will not be facing East or South. I would also recommend placing the bathroom totally separate from the toilet. Why have the toilet in the same place as the bathtub, medicine cabinet and the washbasin? Yes, hang the towels right above the toilet bowl where the after-flush spray will get it! It is disgusting.

I think showers are more hygienic than bathtubs. Sitting in your own dirty bathwater seems risky, particularly when the same water is going everywhere... everywhere on your body!

I am aware that those who use them will build up resistance to the intermingling microbial flora but...

The other thing that horrifies me is the use of toilet paper. Stop it. You are not 100% clean with it. Get a toilet with a bidet attachment. The water gun or sprayer is the best invention since indoor plumbing!

Septic tank

Vāstu experts today recommend the septic tank (sewage percolation tank) in the North or the East. I would add that that it be as far away from the well, preferably in a place where it would get sunlight all throughout the day to aid in its function as a mini-sewage treatment plant. Shadows of the house, trees, or other structures falling on the tank would limit evaporation and microbial activity necessary to treat the sewage. The tank should also be accessible from the street so that it can be completely emptied once in a few years. (Otherwise, built-up chlorides and other salts start to leach into the ground water and make it hard.)

Parking area on a plot

As the North and East are exposed to the Sun and air, they are recommended for parking areas. The North-East corner is not recommended for a shed.

Basement

For residences, Vastu Shastra does not recommend building dwelling place in a basement. In modern apartment complexes, the ground floor is used as parking space. This is fine but no dwelling units are recommended there.

Ambiguous/conflicting interpretations

Sometimes, different Vāstu authors prescribe exactly opposite recommendations. In South India, the practice is to place doors and windows placed in a straight line. You can stand in the main entrance and see what is going on in the backyard if all the doors in between are open.

Some Vāstu authors say it should not be so as to preserve the "energy" within the house. Perfectly aligned windows and doors make it easy to spot human intruders and animals that may decide to wander in. There are practical benefits from the custom.

If there is a conflict, you should follow the local custom. If you are unable place doors aligned from front to back, you can use windows aligned like that. In Tamil Nadu, a small triangular hole is chipped into the wall to meet this requirement.

The nadumuttam in nalukettu houses are usually depressed. Some Vāstu authors say Brahmasthal should be at higher than other places. I doubt it. It should be lower to avoid flooding.

Namboothiris, the latter-day Brahmans of Kerala, say they follow Vāstu but the building plans of old Namboothiri *manas* that I have seen were not even square or rectangle. They were L- and T-shaped!

Some plots are at the end of a cul-de-sac. This is considered inauspicious. Many of these house-owners build a mini-Ganesha shrine on the wall and go on with their lives. That is the most prevalent Vastu remedy in South India for a cul-de-sac.

The reason I suggest deferring to local customs or traditions rather than Vāstu *per se* is because the former tends to be more evolved and adapted to the region. Local customs and traditions also tend to be light on the environment... sustainability, 'green' technology, all that jazz.

Non-Vāstu considerations

Apart from Vāstu, there are other things that matter for the life of the property and for the dwellers' peace and prosperity:

- Belief and worship of Gods and Goddesses.
- Dharma (your deeds and charity).
- Living modestly and avoiding ostentatiousness - do not evoke spite or jealousy in others.
- Use natural building and painting materials, as much as possible.
- Consider whether Western thrones are the option for you. I am not a microbiologist but I think Asian-style squat potties are safer. (Note: This is my own illustration (created as a joke), not US CDC's stand.)

- Water is the ultimate solvent. Given enough time, it will destroy everything in its course. Water sources such as taps, sinks and washbasins should not be placed near doors or windows. It will weaken these structures. Place them where sunlight can dry them quickly. Pipes should be used in terrace roofs and sunshades to evacuate water as quickly as possible. Even walls need to be protected from water. The house should be regularly whitewashed and cracks should not be left unattended. Before rains, all pathways for evacuating water should be cleared of obstructions such as

dried leaves and twigs.
- Ensure that plumbing lines short and simple. All nodes in a plumbing line should be easily accessible, maintainable and repairable. For example, a valve near tap will make it easy replace the tap without emptying the pipe and the tank.
- Do not create any area that limits cleaning/maintenance activities or becomes a haven for pests.
- When installing AC wiring, make provisions for or install additional conduits for inverter wire, telephone wire, computer networking cables, CCTV cables, DC supply wires, etc.

- When installing the roof, make provisions for future installation of solar panels.
- Acquire enough tools so that ANYONE including children, elderly and womenfolk can do most repair/maintenance jobs. Leave complicated and dangerous works to experts.
- Pursue healthy living choices - exercise; regular, fresh and home-cooked meals; avoiding processed food and artificial ingredients/additives; using metal/ceramic containers, instead of plastic; natural clothing suitable for local climate, traditional cosmetics and cures; resorting to allopathic medical interventions only as a last resort, after exhausting all other options; avoiding/mitigating pollution.
- Limit the amount of furniture and other articles in the house. Excess furniture creates the illusion of the crampedness. Limiting them creates the illusion of spaciousness.
- Careful planning, regular maintenance, plain common sense and presence of mind is recommended for all.

I think the information in these chapters is enough for you to create your house plan. In the last chapter, I have provided several sample house plans implementing these recommendations. When your plan is ready, take it to a good Vāstu consultant and a civil engineer for their revisions.

Sample Vastu plans

In this chapter, I have provided plans for a few tiny houses to illustrate Vastu concepts. For practical implementation, I have also provided plans for similar but more spacious houses of ~1200 sqft. These are tiled *naalukettu* houses that include the toilet inside the plan. For this reason, they are not Vāstu-perfect. These days however we prefer indoor plumbing as a more-than-acceptable compromise.

For purists though, the tiny house plans that incorporate the toilet and even the shower on the outside are Vastu-perfect. In these plans, the two conveniences are in the *Varun* (Rain) position or the *Vayu* (Air) quadrant. The area enclosing the toilet and shower is attached to the house by a different type of perimeter wall, foundation and roof so as to the make the main plan Vastu-perfect. These plans also require an additional door opening to the outside.

House plans for East-facing plot

An East-facing house is the best. In such a house, it is auspicious to have the main entrance in the 6th and 7th Vāstu Mandala positions on the East wall. That is why the hall (containing the main entrance) takes the North-East corner of this illustrative tiny-house plan. It accommodates four bedrooms!

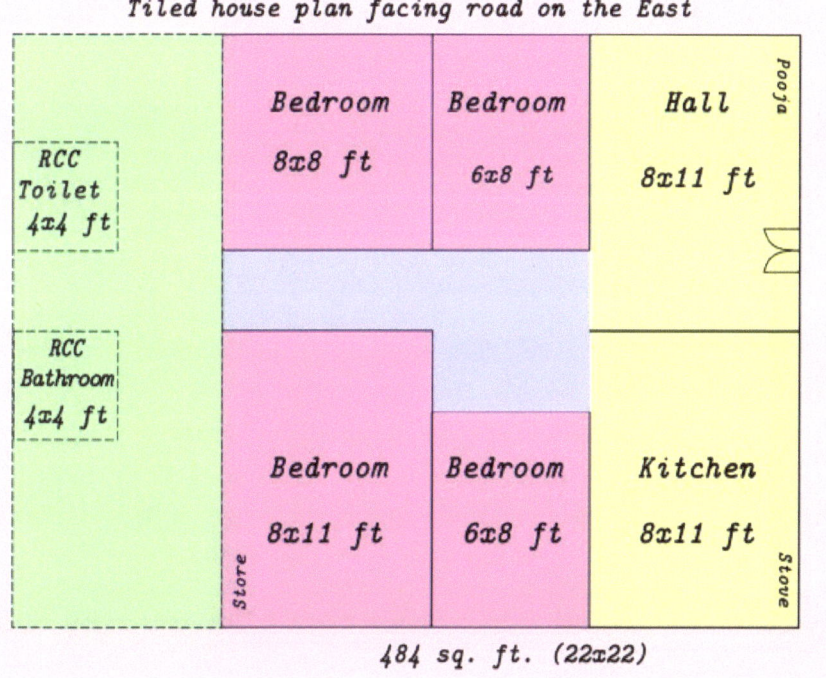

The room with the South-West corner is ideal for the master bedroom. The room with the South-East corner is ideal for the kitchen. The centre is left empty, as required in all Vastu-compliant houses. The pooja place should be a covered one, particularly as it is in the hall. The idols should face East.

The toilet and bathroom are outside the house on the West. In case you are wondering, bathtubs are rare in India. ≥ 💡 ≤ A typical Indian bathroom has a bucket or a water tap or a showerhead as the source of water.

The next plan is a *naalukettu* version of a similar East-facing house. This plan is for a more practical 1200 sqft house. The 20x10 sqft hall is the biggest room. Most other rooms are 15 feet on the longest side. Two tiny bedrooms are 10x10 sqft.

The centre known as the *nadumuttam* is open to the sky. Rather than having a verandah around the house (to beat the heat from the Sun), these houses have an inner verandah. This also improves air circulation inside the surrounding rooms.

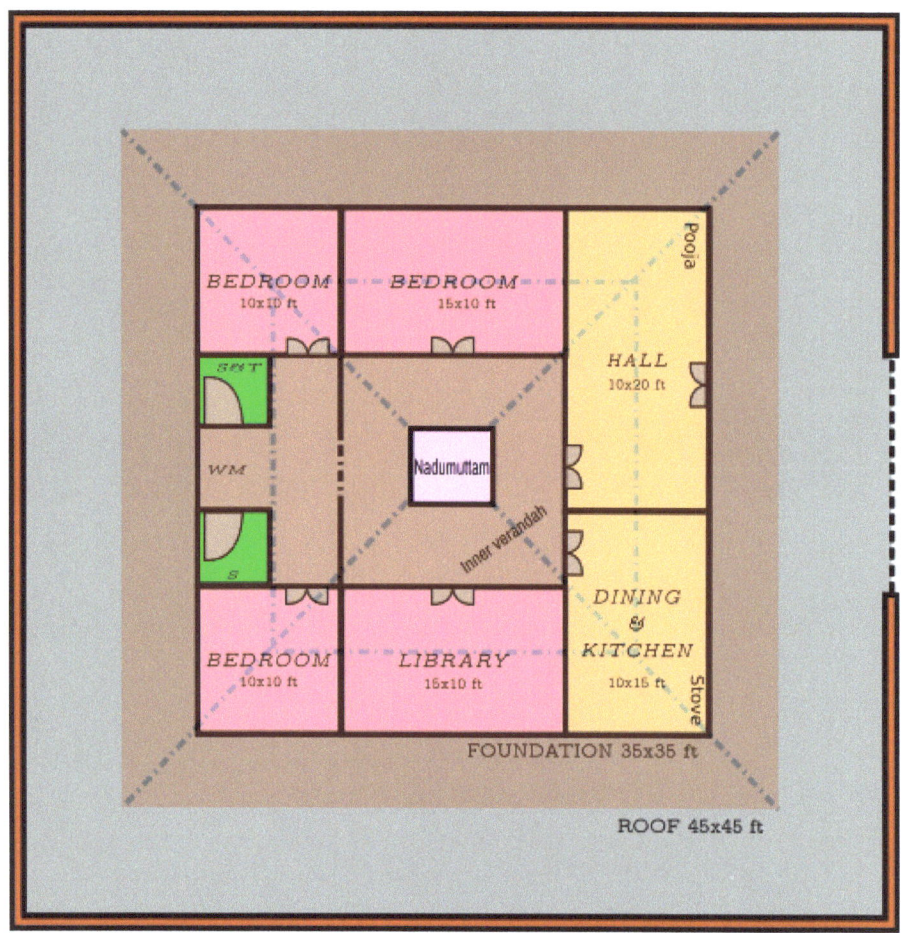

EAST-FACING PLOT 60x60 ft

The toilet and showers are inside the house. For this reason, this plan is not 100% Vastu compliant. However, the two convenience are positioned to take advantage of the North-West *Vayu* (air) quadrant and the *Varun* (rain) position.

House plans for North-facing plot

The next plan is for a 3-BHK house. As is the norm, the main entrance is on the right half of the North wall facing the roadside. The room with the North-East corner is designated as a library, as the *pooja* place should ideally NOT be in a bedroom. If required, it can double up as a *quasi* bedroom.

Tiled house plan for plot facing road on the North

RCC Toilet 4x4 ft	Hall 11x8 ft	Library 11x8 ft / Pooja
RCC Shower 4x4 ft	Bedroom 8x11 ft / Store	Bedroom 8x6 ft	Kitchen 8x11 ft / Stove

484 sq. ft. (22x22)

The room with the South-West corner is the master bedroom. The room with the South-East corner is used for the kitchen. A tiny bedroom has been squeezed in between these two rooms.

The *naalukettu* version of the above plan is a 4-BHK plan. There are two tiny bedrooms (10x10 sqft), one master bedroom (15x10 sqft) and one library (15x10 sqft). The centre in the tiny *nadumuttam* (5x5 sqft) is open to the sky. The inner verdandah is five feet wide. The roof extends outside the foundation to by five feet. The 35x35 sqft house is centered in a 60x60 sqft plot.

The toilet is in the *Vayu* (air) quadrant. It also has a shower and ideally placed to be used by senior members of the family. The shower is in the *Varun* (water) position of the Vastu mandala. As the toilet is inside the house, this plan is not 100% Vastu compliant.

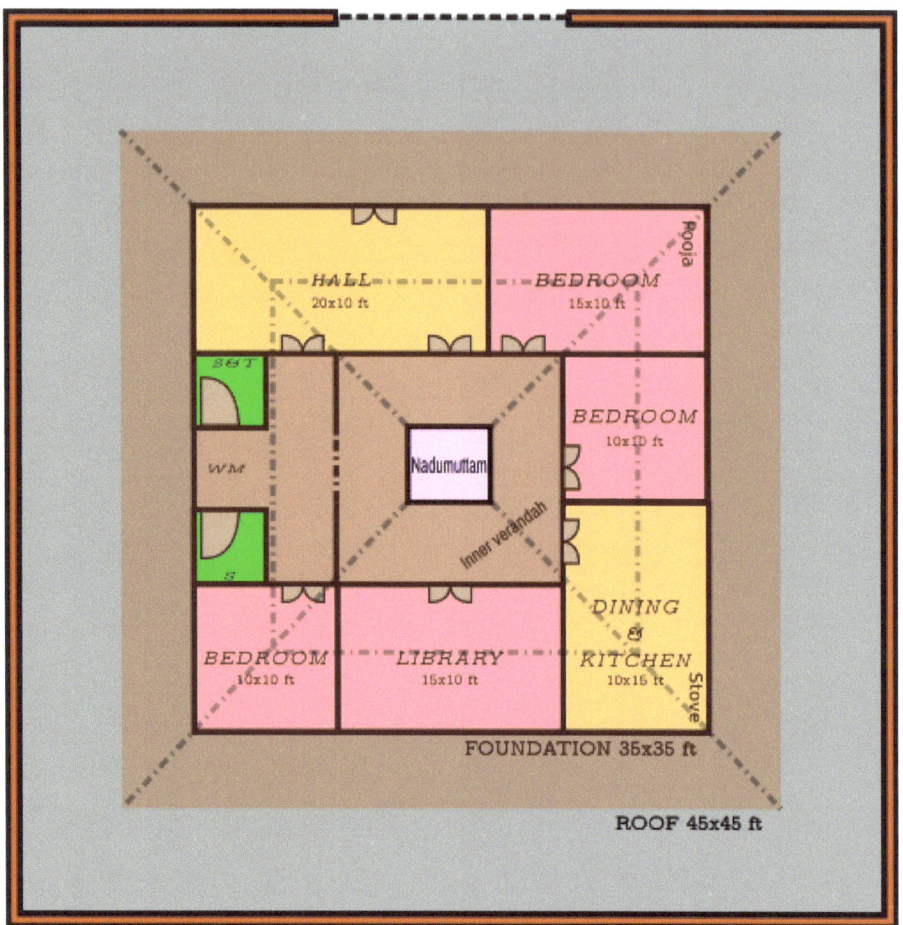

NORTH-FACING PLOT 60x60 ft

House plans for West-facing plot

The next plan is for a 3-BHK West-facing house. The hall is in the West and the kitchen is in the East. Main entrance is in the right side of the West wall and left side of the hall. The centre is empty. All rooms open into it.

Because of the extremely small size of the house, the South-West quadrant cannot be used for the master bedroom. The North-West quadrant can be used for the master bedroom. The hall with the main entrance has greater precedence. The library in the North-East quadrant can also be used as a bedroom if space is a constraint.

The toilet and bathroom are outside the house but attached to the house on the North (represented by Air and Water) with a different type of perimeter wall, roof and foundation.

Tiled house plan in plot facing road on the West

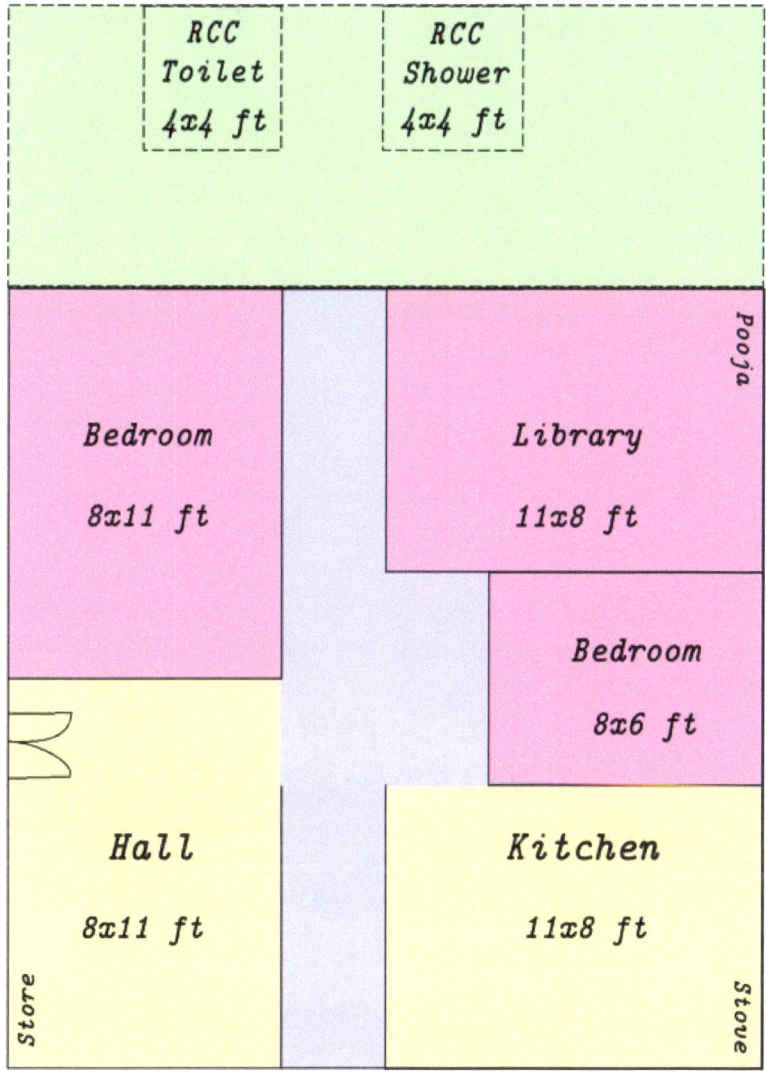

484 sq. ft. (22x22)

In the *naalukettu* version of the above plan, the bedrooms are in the Northern half for no particular reason other than that the hall and kitchen had to be in the Southern half.

The toilet and shower are in the South. Even though the toilet is on the same half as the kitchen, the *naalukettu* configuration ensures that their entrances are as far away as possible and there is no direct wind draft. The shower and the toilet are on either side of the *Yama* (death) position. However, they should not positioned in the Yama position. In fact, they should never be placed in the middle of any side. The space between the shower and toilet in all of these naalukettu plans can be used for something insignificant such as a washing machine.

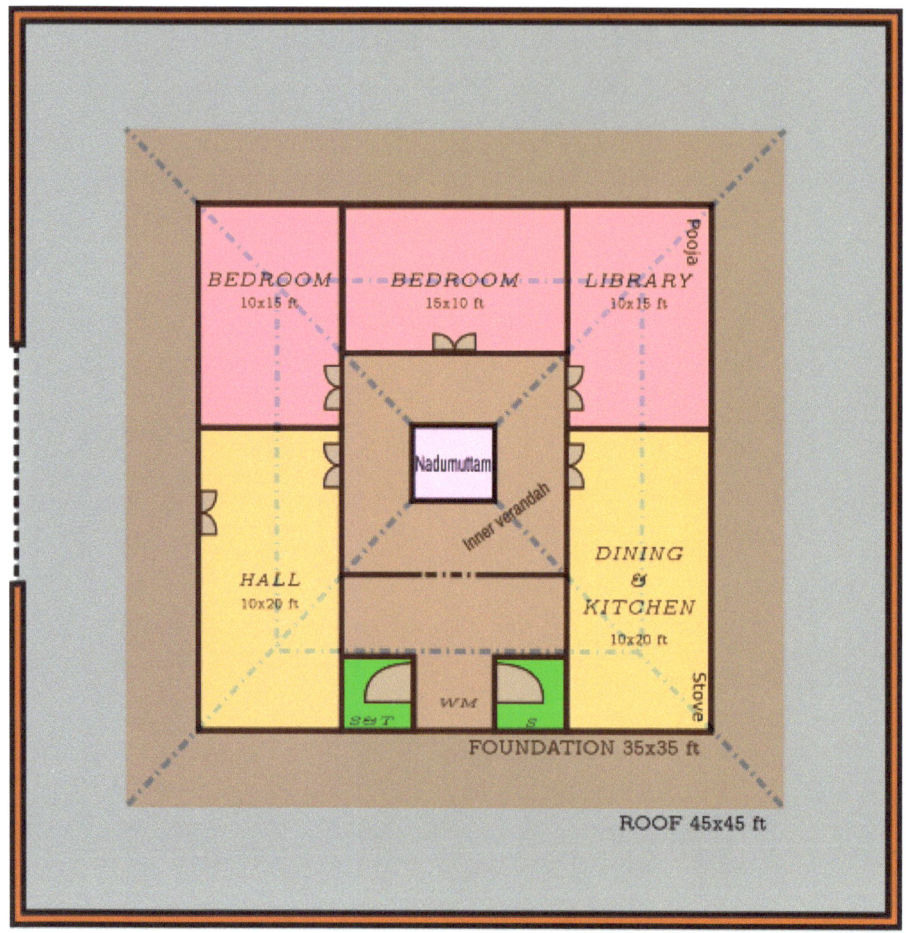

WEST-FACING PLOT 60x60 ft

House plans for South-facing plot

A South-facing plot is permitted in Vāstu Śastra. The fact that this direction is ruled by Lord of Death *Yama* is not a problem, as South is a cardinal direction. It has equal weight as the other three directions.

However, a plan in this direction is difficult to develop if you cannot build more than two rooms in the front. The kitchen's place taking the usual South-East corner is lost to the hall because the main entrance has to be in the right half of the South wall. In this plan, the kitchen is placed in the North-East *Jal* (water) quadrant, even though some Vastu experts consider the North-West *Vayu* (air) quadrant as more compatible with the usual South-East *Agni* (fire) quadrant.

In olden days, wood and coal was placed in the kitchen and the North-East quadrant ruled by *Jala* (water) was not advised. Today, we use a different fuel

(CNG) so the quadrant is not prohibited anymore. It is more important for the kitchen to be on the East side. If the kitchen was placed in the North-West quadrant, then the stove would have to be in South-East corner of that room. This is less than ideal as it would be close to the centre of the house. For these reasons, the North-East quadrant is a better alternative for the kitchen.

The toilet and bathroom are outside the house but attached to the house on the North (represented by Air and Water) with a different type of perimeter wall, roof and foundation.

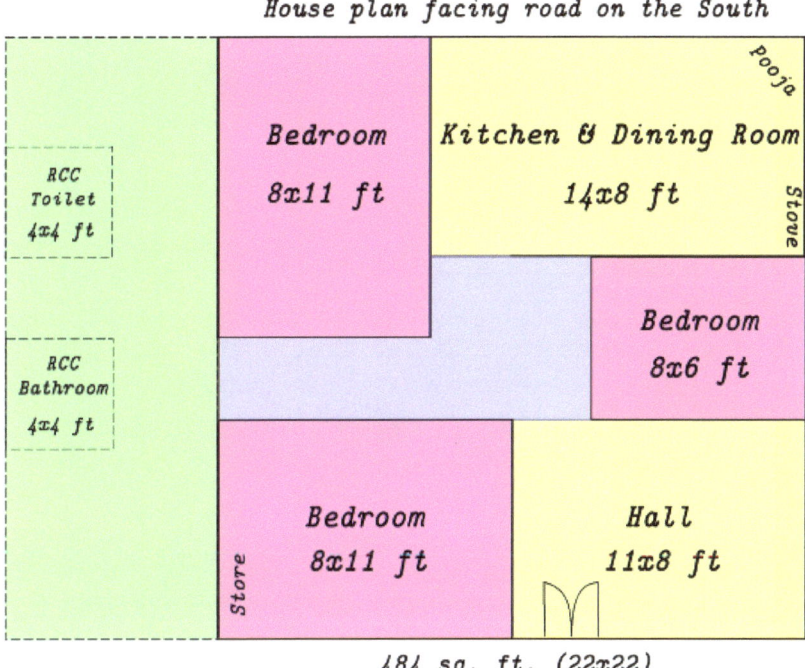

House plan facing road on the South

484 sq. ft. (22x22)

The *naalukettu* version of the above plan is a 4-BHK house. Because of the longer South wall, it is possible to have three rooms in on the front wall. Thus, the hall with the main entrance can be in the South. The kitchen can also be accommodated to take the traditional South-East corner.

An inconvenience with this plan (unlike the other three) is that the bedrooms on the West side do not open into the inner verandah. This does not necessarily make them non-compliant with Vastu. These bedrooms can be used for guests or those who need a quiet place away from the hustle and bustle of the other rooms. One of them, the one in the North-West, is ideal for senior citizens who need quick access to the toilet.

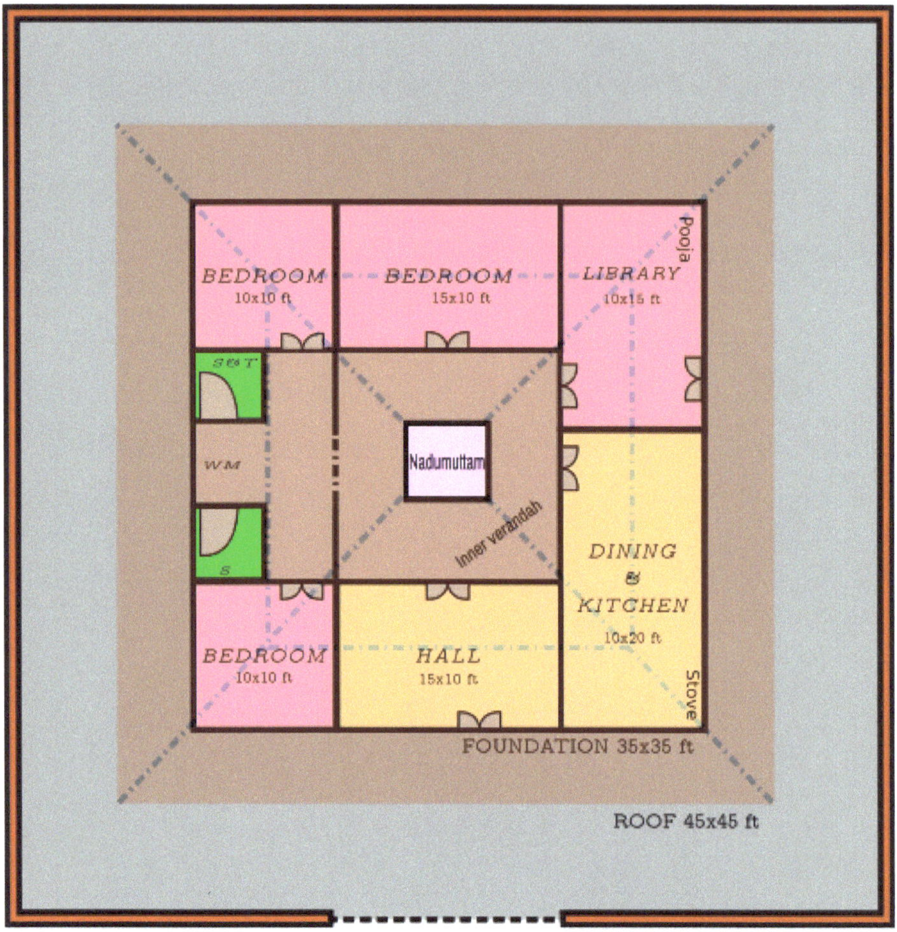

[Floor plan: SOUTH-FACING PLOT 60x60 ft, FOUNDATION 35x35 ft, ROOF 45x45 ft, with Bedroom 10x10 ft, Bedroom 15x10 ft, Library 10x15 ft, Pooja, Nadumuttam, Inner verandah, Bedroom 10x10 ft, Hall 15x10 ft, Dining & Kitchen 10x20 ft, Stove, W.M, S & T, S]

Note on *naalukettu* houses

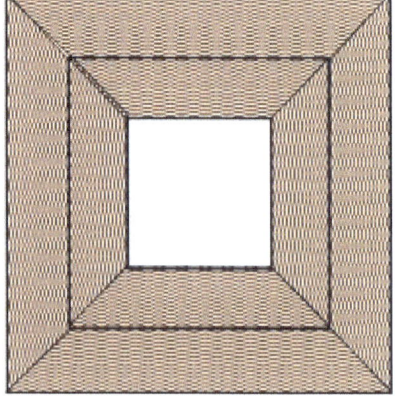

Kerala is famous for its tiled houses (*ottuveedu*). The coziest ones have an attic (*thattil*). A few are four-sided houses (*naalukettu*) with an open-to-the-sky central courtyard (*nadumuttam*).

A *naalukettu veedu* has rooms all around an open central courtyard (*nadumuttam*).

The temperature in the inner verandah is much cooler than outside. Underground pipes are used to evacuate rain water that falls in to the courtyard.

Many Vāstu authors and consultants do not recommend anything within the Brahmasthal (the central *nadumuttam*), not even a depressed floor there. However, I have seen houses with a well and *thulasi thara* within the central courtyard. I am not sure. Follow local traditions. Very few houses, even old traditional *manas*, are 100% Vāstu-compliant anyway.

Two-storey tiled house with full-height thattil (upper level)

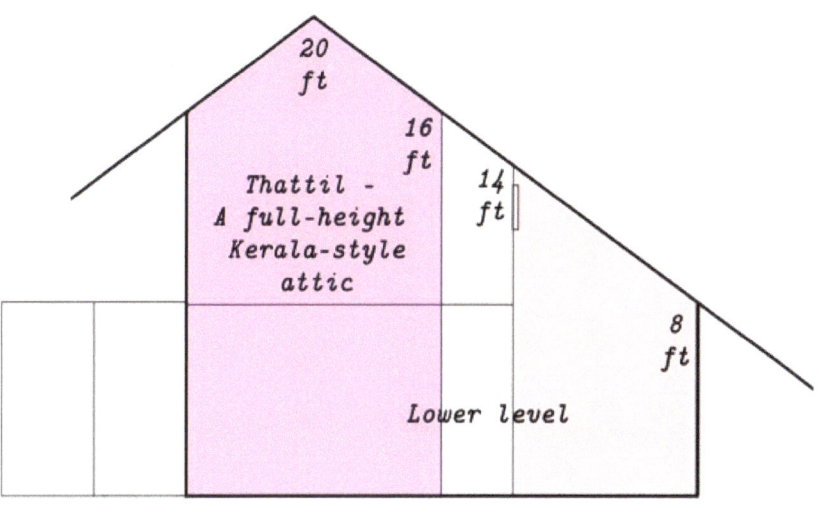

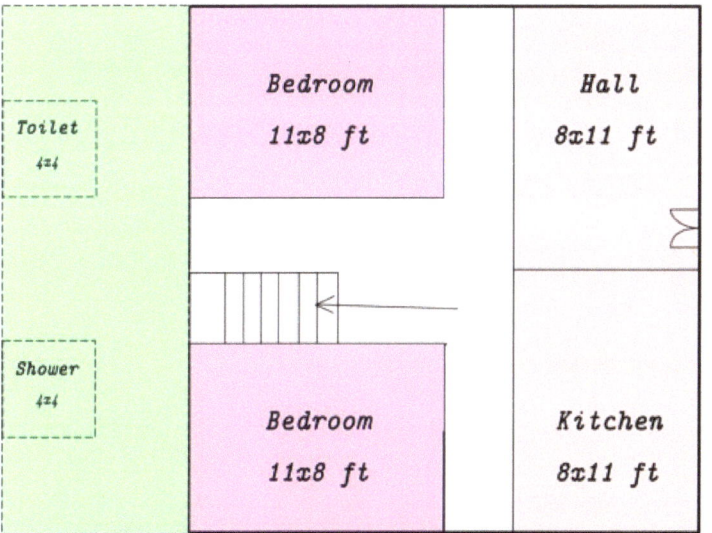

This 484 sqft 4-BHK plan is for a typical Kerala-style tiled house. There are four big bed rooms within an area of 484 square feet. It has a thattil or upper level that is full height. The thattil is reached by a staircase. A deviation from the traditional style is the contiguous roof that stretches from the West to the eastern side. In the traditional style, the eastern half would have a separate roof at a lower level. (Leaves, dirt and other debris accumulate where the two roofs meet. Without maintenance, there will be waterlogging.) The contiguous roof has been made to permit a balcony that would connect to the eastern side.

The upper level is on the western side. The eastern side does not have an upper level. Windows above the regular roof level will have to be made in the western walls of the hall and kitchen to enable communication from the upper level. The verandahs make the plan extremely spacious.

North-facing plot

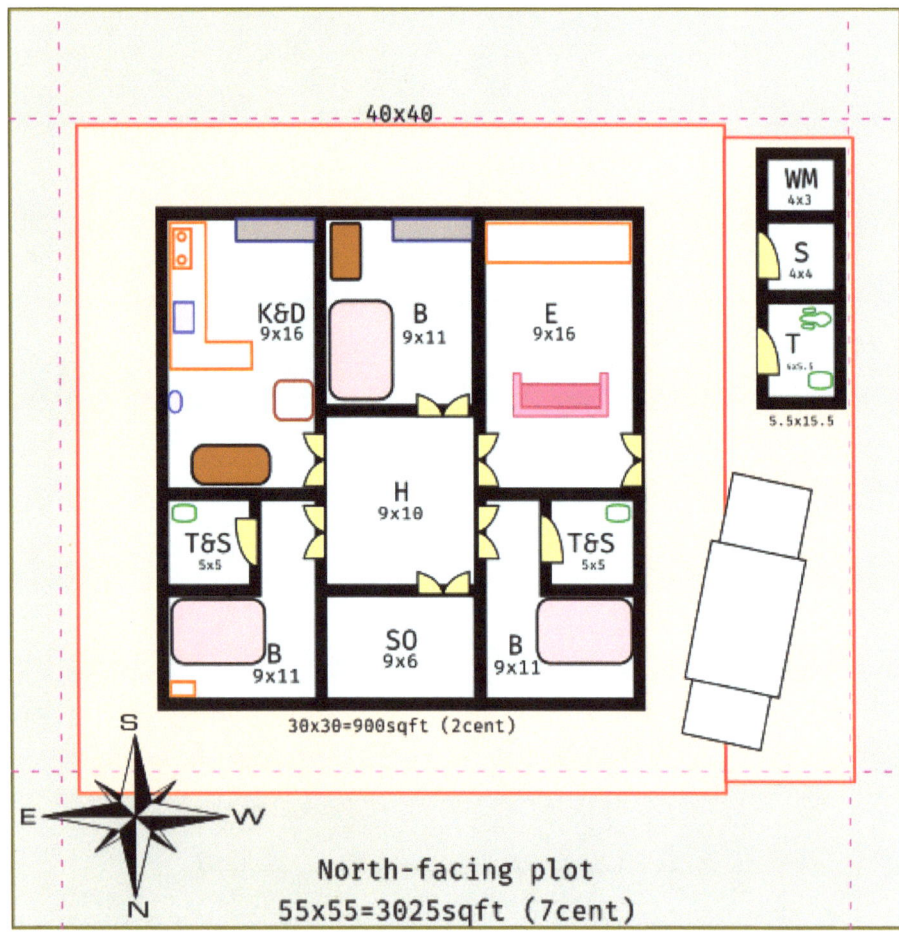

This is a 7-cent plot with a 900sqft house. Open space around the house is three metres in the front, two metres in the rear and one metre on the sides, as per regulations in Kerala. There is a sitout in the front of the house. (It does not have a front wall.)

As the centre of the house is to be left empty, it is specified as the hall. It can be furnished with just a couple of sofas and a coffee table. This plan accommodates three bedrooms and a home theatre. Two of the bedrooms have attached shower and toilet. One bedroom (the one in the North) does not have this feature. The kitchen is as long as the home theatre, as there is never enough there.

There is an additional door (in the home theatre) for entry from and exit to the outside. There is a common bathroom and toilet outside the house that can be accessed using this door. Next to them is the laundry room with the washing machine. These three rooms are two metres offset from the house.

South-facing plot

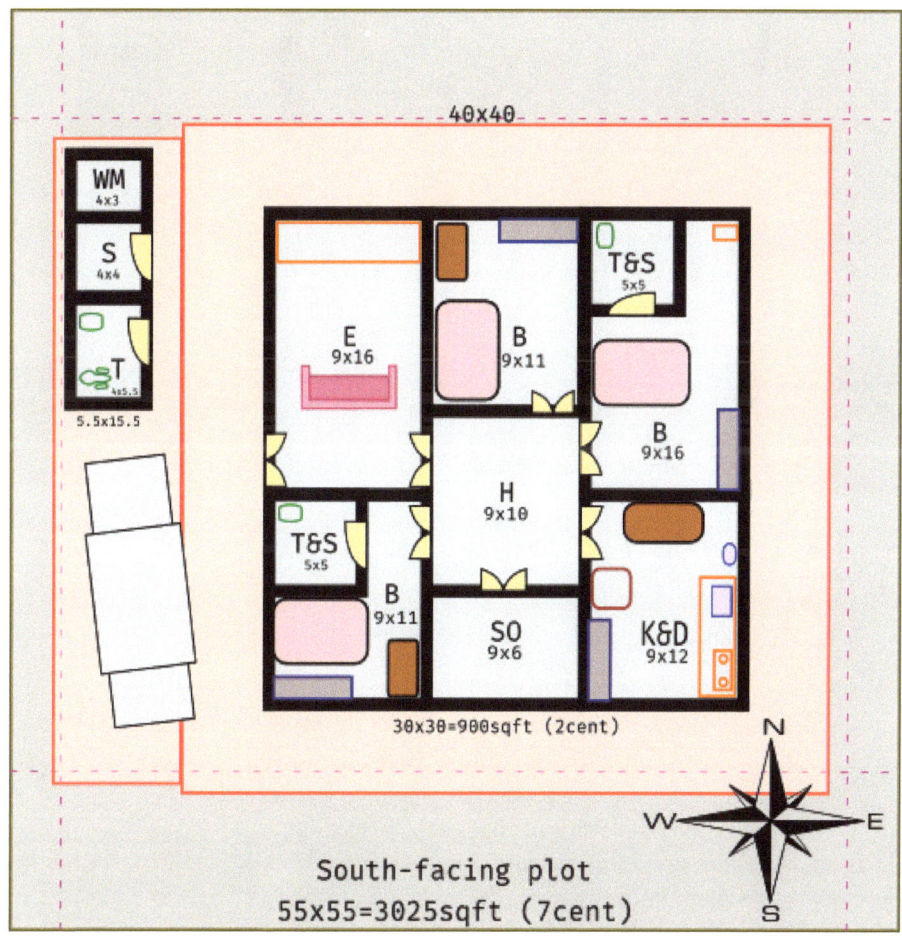

This plan is similar to the previous one but it faces South. The main entrance is in the 5th quadrant (Yama) as prescribed in the *Vastu Sasthra* of *Matsya Purana* so let there be no worries. The other option is to have the main doorway in the 7th quadrant, which is not available in this plan. The kitchen is not as long as the home theatre but the master bedroom is.

East-facing plot

This plan also accommodates a master bedroom.

East-facing plot
55x55=3025sqft (7cent)

You may note certain common features in these (last four) plans:
- there is a sloping roof on the house. There are sloping sunshades all around the house. They project five feet away from the outer walls of the house, protecting it from the Sun and rain.
- all rooms open in to the hall.
- the kitchen has space for a stove, a sink, a fridge, a cupboard, a dining table and a wash basin.

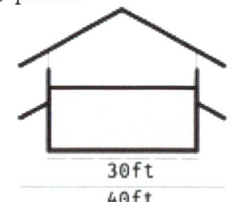

- the outside bathroom and toilet are not on the side of the kitchen.
- the South-West corner is pinned down by a cupboard or a shelf. (The foot of a bed can also be positioned there.)
- there is space to park the longest mass-produced car (in India) within the compound and under a shade.

West-facing plot

Well, you have finished the book. If you give it a good review or rating (★ ☆ ☆ ☆ ☆) online, it would be much appreciated. If you have any corrections or suggestions, write to me at **Info@VSubhash.Com**.

Annexures: Magnetic declination maps

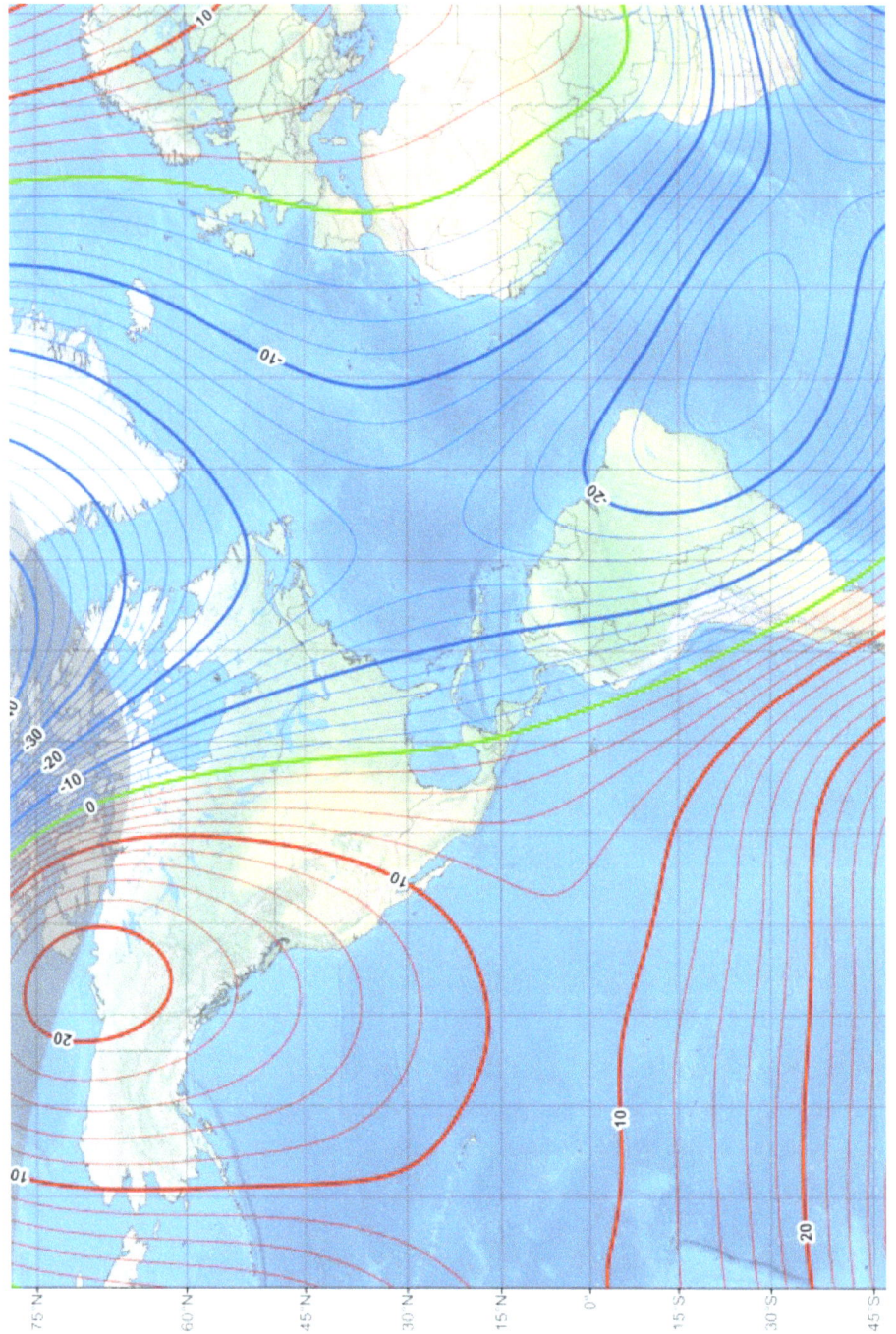

In these maps from the US National Oceanic and Atmospheric Administration

(NOAA), green lines have zero declination, red lines show declination to the West and blue lines show declination to the East.

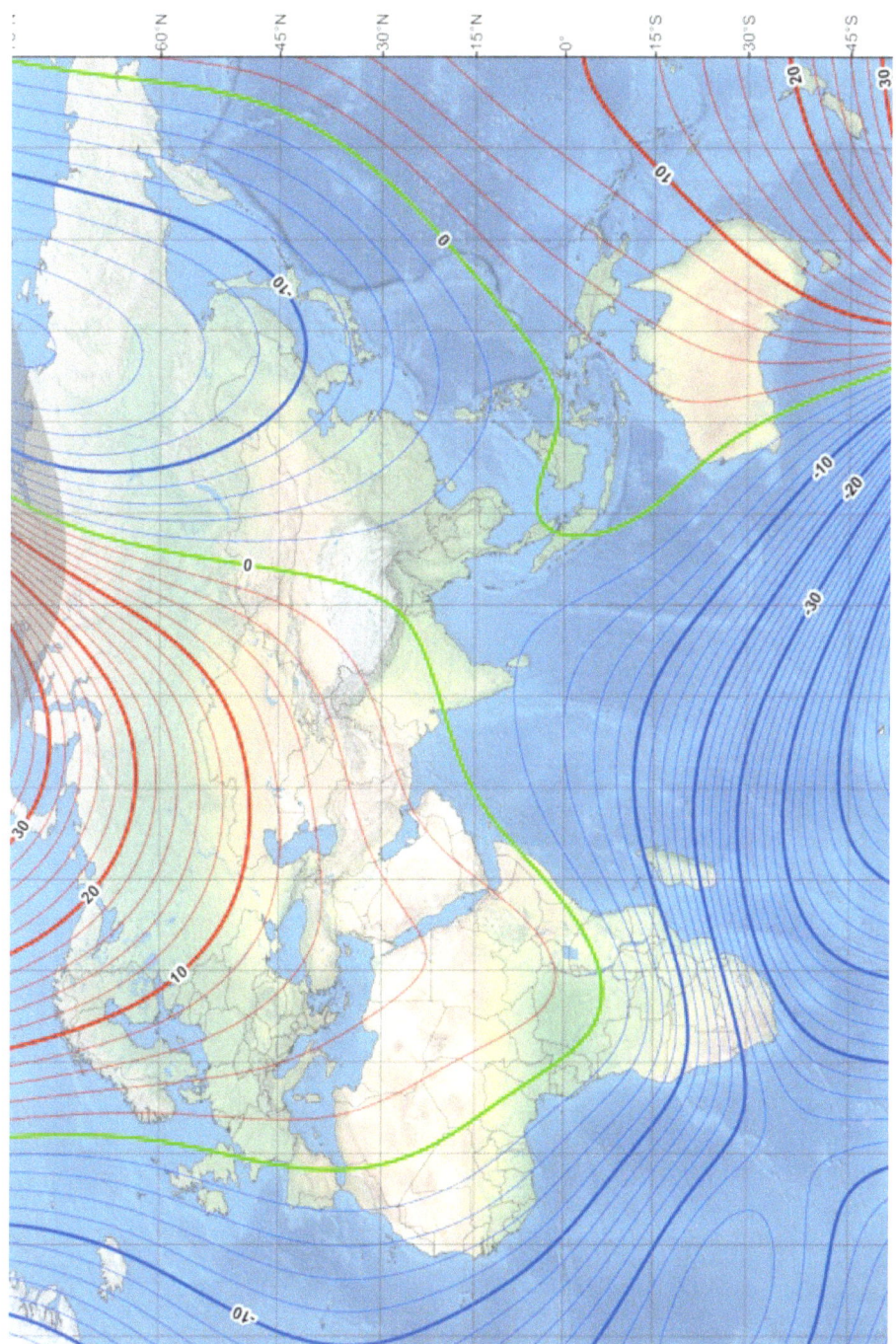

Books By V. Subhash

I invite you to visit my site **WWW.VSUBHASH.IN**, and check out my other books, special discounts, sample PDFs and full ebooks. In 2020, I started publishing books. For two decades before that, I have been publishing feature articles, free ebooks (old editions still available), software (server/desktop/mobile), reviews (books, films, music and travel), funny memes and cartoons. You can follow these adventures on my blog: **http://www.vsubhash.in/blogs/blog/index.html**

My books for children are under the pseudonym **Ólafía L. Óla** (because it has laugh and LOL).

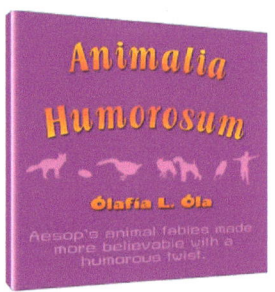 

 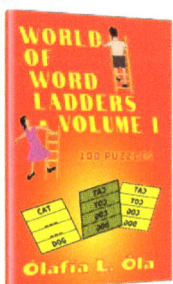 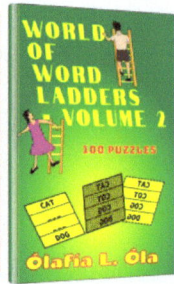 

 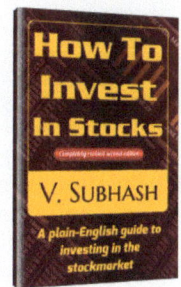
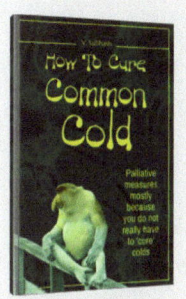  

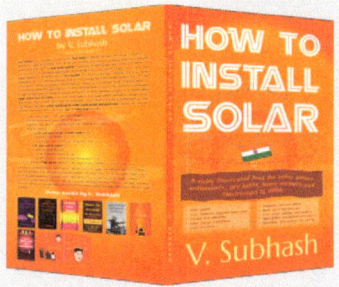 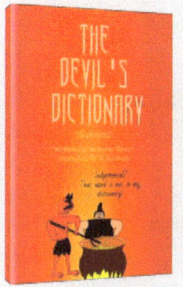 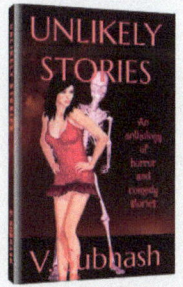

About the author

You will never find another guy with so many talents as this one! V. Subhash is an invisible Indian writer, programmer, cartoonist and humourist. He grew up in Chennai but is now settled in his native Kerala. In 2020, he published one of the biggest jokebooks of all time — *2020 Fresh Clean Jokes For Everyone*. Subhash was inspired to write it after years of listening to vintage American radio shows such as *Fibber & Molly* and *Duffy's Tavern*. In the same year, he followed up with a how-to book on the multimedia software FFmpeg and a 400-page volume of 149 political cartoons. How did he do that? Subhash pursues numerous hobbies and interests that inevitably became the subject of his books — like *Cool Electronic Projects*, *How To Invest In Stocks* and *How To Install Solar*. He used to have pet tortoises but they died in a parking accident. Everyone was crushed. For two decades before 2020, Subhash used his personal website **www.VSUBHASH.in** as the main outlet for his writing while also accumulating a lot of unpublished material. By 2022, he had exhausted all that he could publish. Meanwhile, his probe into 'aliens' had revealed that they are just ordinary employees/contractors of US military and space agencies. They begged him not to write anything so he published his findings in his debut fiction title named *Unlikely Stories*. The stories turned out to be supernatural/paranormal/sci-fi fantasies with ample doses of action, horror and humour. In 2023, Apress (SpringerNature) published his rewritten and updated FFmpeg book as *Quick Start Guide To FFmpeg*.

www.ingramcontent.com/pod-product-compliance
Lightning Source LLC
Chambersburg PA
CBHW040252220526
45473CB00001B/450